Green your life

Green your Skool

Mariam Karim-Ahlawat

Bharat Shekhar

teri

The Energy and Resources Institute

An imprint of The Energy and Resources Institute

First published in 2014 by
The Energy and Resources Institute
TERI Press
Darbari Seth Block, IHC Complex, Lodhi Road, New Delhi 110 003, India
Tel. 2468 2100/4150 4900, Fax: 2468 2144/2468 2145
India +91 ■ Delhi (0)11
Email: teripress@teri.res.in ■ Website: http://bookstore.teriin.org

ISBN 978-81-7993-492-0

Publishing Head: Anupama Jauhry
Editorial and Production Teams: Ekta Sharma, Himanshi Sharma;
Aman Sachdeva
Design and Illustration Team: Santosh Gautam, Vijay Kumar;
Yatindra Kumar, Vijay Nipane
Image Researcher: Shilpa Mohan

Printed and bound in India

This book is printed on recycled paper.

Contents

AT THE SCHOOL GATE

Cycle to school if you can. It saves energy, doesn't pollute, and keeps you fit

Vehicles should be well maintained. Exhaust fumes contain harmful chemicals

Vehicles should be kept away from schools to avoid pollution and noise

CANTEEN

Use medicated handwash to wash your hands!

POPU'S BLOOPERS
• Asking for junk food like fries and soda
• Making fun of a poor student
• Throwing tantrums

Don't serve food with your hands; use spatula or tongs instead
Don't reuse oil after frying. Not good for health!
Use cloth wipes – as they can be reused – to clean counters
Steaming hot, germ free!
Fruit and vegetables for good health!
Cloth caps to cover hair
My homemade desserts! Especially good for kids!
Helping elders in their daily chores is a good habit
Spot Popu's bloopers
A poor goat girl in our canteen?
In a great country, everyone has the right to education!
No sodas? No plastic gloves? No fries? No AC?
Sodas are bad! Plastic creates waste! Fried food is junk! ACs harm the environment!
Eat fresh, local dishes!
Chairs from recycled plastic
6-7

OUR GREEN SCHOOL

Use harvested rainwater for flushing toilets

Electricity from solar energy

Green buildings help save water and energy.
WATER HARVEST
Solar energy based sprinkler system
Our school makes its own electricity with solar panels!
We collect rainwater for REUSE!
All our school projects make the world a better place to live!
More oxygen better health Grow lots of trees for shade
Old paper makes beautiful things! Reduce, Reuse, and Recycle!
We grow lots of trees for shade.
Papier-mâché pots from old, waste paper
POPU'S BLOOPERS
Throwing plastic in the wrong bin

IN THE CLASSROOM

SPOT POPU'S BLOOPERS
* Asking for AC when there is a ceiling fan and large windows to keep room cool

Air-conditioners harm the ozone layer. Ozone protects us from harmful UV rays
Gift or reward students with potted plants
SHAPES
Decorate empty boxes to make nice pen holders and chalk boxes
Dust Bin
Use cartons to store items like pencils, erasers, sharpeners, paints and brushes.
Don't throw tetrapacks. Reuse them creatively to form decorative items like pen stand
Multipurpose
Recyclable Paper
10-11

IN THE
PLAYGROUND
Vegetables grown in your backyard are healthy and tasty!
But I don't feel like playing.
When you have free time, play till the bell will chime.
Root out weeds as they choke other plants
We will build our new science wing here. Who needs a playground?
Green bins for organic waste
Our own Garden of Eden, but it has weeds.
Playtime is as important as studies are!
Water, water everywhere, but be careful about what you drink!
Do not drink water immediately after playing. Cool down a bit first
Spot Popu's bloopers
I'll keep you in this box!
Hey, don't trap the butterfly!
Organic manure is magical for plants
Use filtered water for drinking

SPOT POPU'S BLOOPERS
- Drinking immediately after playing
- Drinking untreated water

IN THE WASHROOM
Get steel mugs as they last longer
Chirr! Chiirr! Time for more monkey mayhem.
Students must keep their school clean
POPU was here
Graffiti, graffiti on the wall. Popu's the smartest of them all!
Spot Popu's bloopers
Fix leaking taps and loose fittings to avoid wastage of water and keep the floors clean.
Throw rubbish in the dustbin
Do not waste toilet paper
SPOT POPU'S BLOOPERS
• Touching the dirty pipe with his hands
• Dirtying the toilet doors of his school

Use medicated soap or liquid to wash hands
Go for waterless urinals. Cleaner, greener technology!
Use cloth towels to wipe hands as they can be reused.
We often waste water while brushing teeth and washing hands. It sets bad example for others
I ape Popu!
Water is precious.
Dangerous substances like acid need to be stored safely
What a stink!
Whoops. That's so slippery.
Keep the floor dry or people will slip
Wet floors are a breeding ground for germs
Throw rubbish in the bin or the washbasin drain will clog
germs free
14-15

AT THE EXHIBITION

Use old cloths to make new things such as rags, mats, and bags
OLD CLOTHES NEW WEAR
Doormats from old, discarded dupattas
Recycle old stuff. Save money!
Wow! Papier-mâché mask!
Decorations from old gift paper
Use old newspapers to make Papier-mâché items
Use old cloth to make diaries and dolls
Friends, make haste. Buy and reuse my waste!
Decorate old diyas to make them new
POPU'S INNOVATIONS
All of them run on battery. How smart!
Battery-operated, electronic toys harm the environment
Spot Popu's bloopers
POPU'S BLOOPERS
• Bringing old, electronic toys

ACTIVITIES

MAKE YOUR OWN ORGANIC PESTICIDE AND

What you need?
- a clean 2-3 litre jug or plastic bottle, like the ones cooking oil comes in • a spray bottle with spray
4 litres liquid • 2 small onions • a large green chilli or jalapeno pepper • a clove of garlic • one third

With the help of an adult, chop the veggies and blend them in a blender. Make sure you don't rub your eyes after touching the chillies or pepper.

Immerse the blended vegetable paste in hot water and let it stay for half an hour, so that the hot flavours will all seep into the water.

All fruit and vegetable waste is biodegrabale and it works as an excellent manure for plants. So throw the leftovers into the compost bin.

Add some diswashing liquid to the liquid that you have just prepared. Using a funnel, pour this mixture inside a spray bottle and fix the nozzle.

KEEP THE PESTS AWAY FROM YOUR PLANTS!

nozzle • a funnel • a piece of thin cloth from an old cotton stole, dupatta, or sari • a pot that can hold
cup dishwashing liquid soap • a blender • 1 litre hot water • an adult to help you

Strain the mixture through the cloth into the jug using the funnel so the liquid does not spill outside the jug.

Now hold the cloth as shown in the picture above and squeeze it gently so that the remaining liquid is squeezed out.

Now that your organic fertilizer is ready, spray it lightly on plants, over and under leaves also. It won't harm your plants!

Observe your plants closely a little later. You will notice that the insects do come near plants to eat them but will hate the taste and go away.

WORDS TO REMEMBER

- **Country** an area of land that is controlled by its own government
- **Dessert** sweet food eaten after the main meal
- **Disinfectant** chemical substance used to kill harmful germs and bacteria
- **Exhaust** the mixture of gases produced by an engine
- **Fertilizer** a chemical substance added to soil to help the growth of plants
- **Garden of Eden** a very beautiful garden, mentioned in the Bible, where Adam and Eve lived
- **Lassi** an Indian beverage made from yogurt or buttermilk
- **Manure** solid waste from farm animals that helps in the growth of plants
- **Noise pollution** loud or unpleasant noise that is caused by automobiles, airplanes, etc.
- **Oxygen** a gas found in air and is necessary for life
- **Solar panel** a large, flat piece of equipment that uses the sun's light or heat to create electricity
- **Sprinkler** a device that is used to spray water on plants or soil
- **Unhygienic** not clean and likely to make one sick